HABITATS OF THE WORLD

RAIN FORESTS

Written and Illustrated by
Sheri Amsel

A LUCAS • EVANS BOOK

RSVP
**RAINTREE
STECK-VAUGHN**
P U B L I S H E R S
The Steck-Vaughn Company

Austin, Texas

For my brother Dr. Scott Amsel

who saves lives in the rain forest

Consultant: James G. Doherty, General Curator, Bronx Zoo, Bronx, New York

Book Design: M 'N O Production Services, Inc.

Library of Congress Cataloging-in-Publication Data

Amsel, Sheri.
   Rain forests / written and illustrated by Sheri Amsel.
     p.   cm. — (Habitats of the world)
   "A Lucas/Evans book."
   Includes index.
   Summary: Discusses the world's rain forests, their importance, plant and animal life, and conservation.
   ISBN 0-8114-6301-X
   1. Rain forests—Juvenile literature.   2. Rain forest fauna—Juvenile literature.
   3. Rain forests plants—Juvenile literature.   4. Rain forest ecology—Juvenile literature. [1. Rain forests. 2. Rain forest animals. 3. Rain forest plants. 4. Rain forest ecology 5. Ecology.]   I. Title.   II. Series: Amsel, Sheri. Habitats of the world.
   QH86.A75   1993
   574.909'52—dc20   92-8790
                                          CIP
                                          AC

Printed and bound in the United States.

1 2 3 4 5 6 7 8 9 0 VH 98 97 96 95 94 93

Table of Contents

RAIN FORESTS OF THE WORLD

You are about to take a journey to a wild and busy place. Here the air clings like a cloud, hot and moist. The smell is thick with rotting leaves and strong, sweet flower scents. Birds screech from overhead. Hidden animals scurry beneath the foliage causing leaves to rustle, as they search among the trees for food. Insects buzz and hum, and ants march in armies. This is a rain forest.

Fuschia

Bromeliad

There are rain forests all over the world and, though they are alike in structure, each of them has many different types of animals and plants. What they have in common is heat, rain, and very tall trees. Rain forest trees reach to the sky to collect as much sunlight as they can, for sunlight feeds the forest. In fact, rain forests can, because of their dense greenery, collect more sunlight than any other type of wild place in the world.

What does the forest do with all that solar energy? Besides growing towering trees and thick ferns that may be 20 feet tall, it grows huge, sweet-smelling flowers, unusual fruits, and strange bromeliads. But most of all, it keeps the rain forest so hot that when the daily rains fall, they evaporate into steam. This keeps the forest dripping with moisture. Therefore most small animals can stay safely hidden in the treetops, eating leaves, fruits, and insects. They need not risk a visit to the ground for water.

6

Philodendron

Morpho
butterfly

Cecropia
tree

Strangler
fig tree

Banana
tree

Cattleya orchid

7

Far below the thick tangle of trees it is dark, wet, and still on the forest floor. Only small shafts of light filter down to dark-loving plants. Dead leaves and fruits pile up on the ground and rot, releasing all their energy back to the forest soil. Nothing is wasted in the rain forest.

The Rain Forests of South America

In the Amazon rain forest of Brazil, a golden lion tamarin scoops insects and other small animals out of a bromeliad for its dinner. A three-toed sloth slowly munches green leaves. Its fur grows a green moss in the wet air. That helps to hide it from hungry meat eaters. It is a lucky thing, for just above the trees a giant harpy eagle soars and looks for an animal to catch with its sharp claws.

Harpy eagle

Three-toed sloth

Golden lion tamarin

The trees are filled with sound and color—a noisy red-green macaw, a toucan, or a spider monkey. On the forest floor, an agouti feeds on an avocado dropped from above. When it has finished, the leftover seed may take root in the rotting leaves, if there is enough light. A tapir nibbles at leaves. Off in the bush a hunting jaguar, one of the biggest predators in the Amazon, sniffs the air for the scent of food. Any animal caught unaware on the forest floor can become a meal for the hungry cat. A butterfly searches for blossoms, and a red-eyed tree frog waits for insects. The rain forest feeds them all.

Toucan

Macaw

Spider monkey

Jaguar

Tapir

Red-eyed
tree frog

Agouti

11

Hummingbird

Ant

Smaller animals are
busy in the forest, too.
Hummingbirds hover
at giant blossoms to feed on the nectar.
While they collect, they carry pollen to another
blossom, fertilizing the flowers as they go. Wasps, bees, and
bats also pollinate flowers as they collect nectar and pollen.

Ants seem to be everywhere. Busily carrying leaves or scraps of food
back to their homes, they make a marching column for miles. As small
as they are, they can be a fierce forest insect and many can inflict a
painful bite or sting.

There are many butterflies in the Amazon rain forest. The brilliant blue morpho butterfly glistens in the soft light. Owl butterflies search for ripened fruit. Painted ladies, monarchs, and heliconids feed on flower nectar and flash their bright colors throughout the forest.

Plants, like everything else in the rain forest, grow to giant sizes. The kapok tree can grow 150 feet high and have huge branching buttresses at the bottom to balance it in the shallow rain forest soil. In these buttresses, animals like boa constrictors find a safe place to hide. The monkey nut tree grows large "cannonball" fruits. The milk tree, when cut, spills out white drinkable sap.

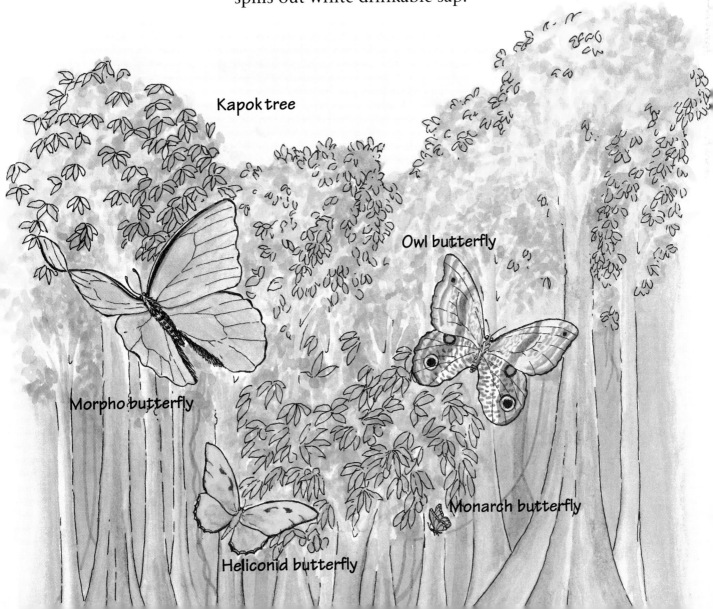

Kapok tree

Owl butterfly

Morpho butterfly

Monarch butterfly

Heliconid butterfly

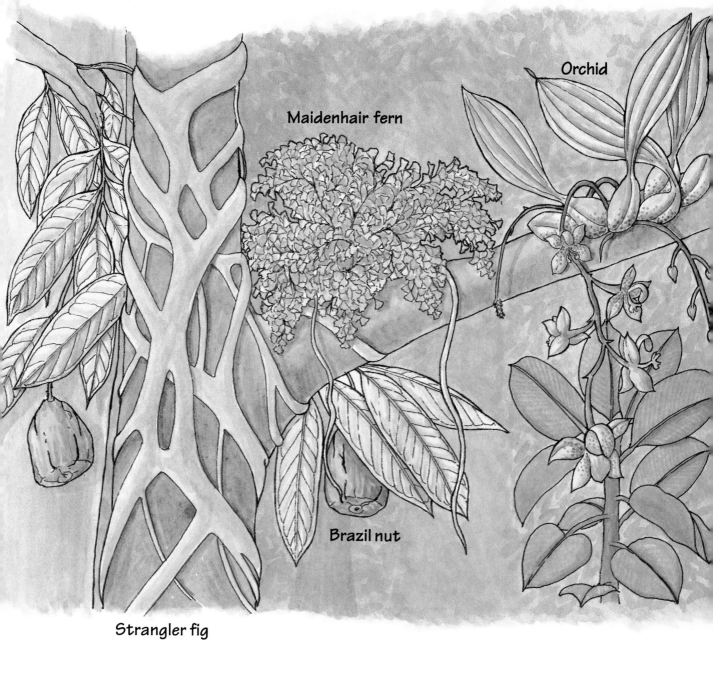

Orchid

Maidenhair fern

Brazil nut

Strangler fig

Brazil nut, mahogany, rubber, palm, and mimosa trees, as well as giant ferns, fill out the forest. Vines called "lianas" wrap around the tree trunks, leaving long trailing tails through the forest. Strangler figs take over trees by rooting in their very bark. Bromeliads, maidenhair ferns, and many colorful orchids root in the crotches of branches where leaves and soil collect.

14

The Rain Forests of Asia

In Asia, Borneo, Sumatra, and the Malay Peninsula, the rain forests look much like the Amazon, but many fascinating differences are found. A giant flying lemur, called the colugo, searches for fruits and flowers. It glides, making a greenish gray flash, from tree to tree.

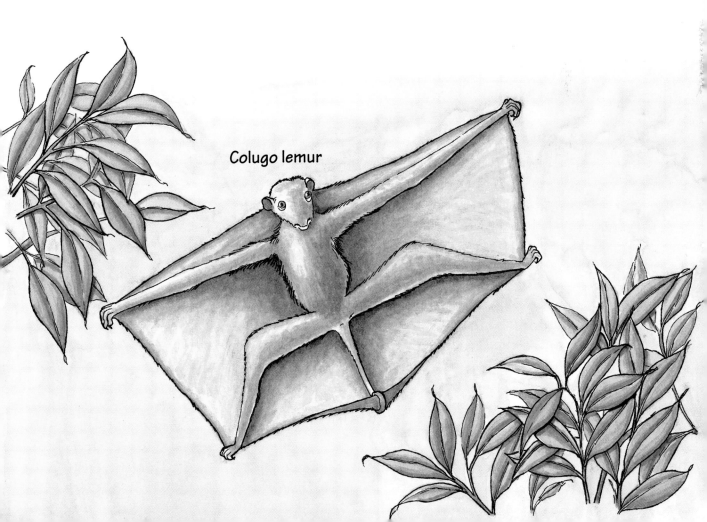

Colugo lemur

Flying lizards also glide through the trees. They don't really fly at all but spread flaps of skin or webbed toes to catch the air. Real fliers, like the hornbill, flap through the forest in search of food.

The forest kingfisher munches on a katydid. Huge orange orangutans lumber from branch to branch, feeding on fruit and stripping leaves with their long fingers and teeth. Gibbons and proboscis monkeys also share the treetops.

Later at night giant fruit-eating bats will awaken to search for ripened fruit. Tigers are the biggest meat eaters here as they search the forest floor for a tapir, tree shrew, or barking deer. Pythons also hunt for unwary animals.

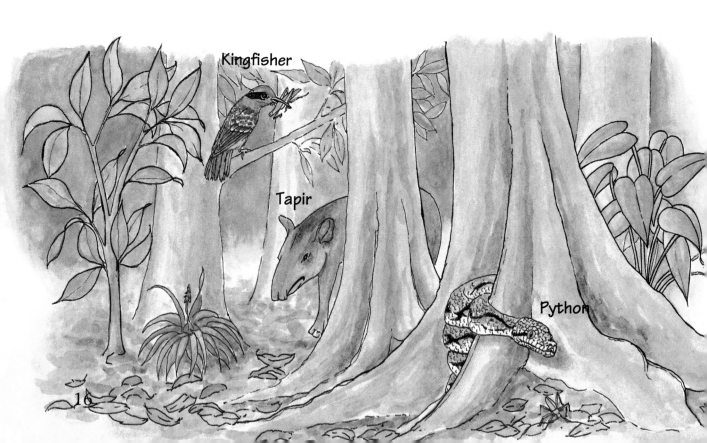

Kingfisher

Tapir

Python

Hornbill

Orangutan

Gibbon

Flying lizard

Tiger

17

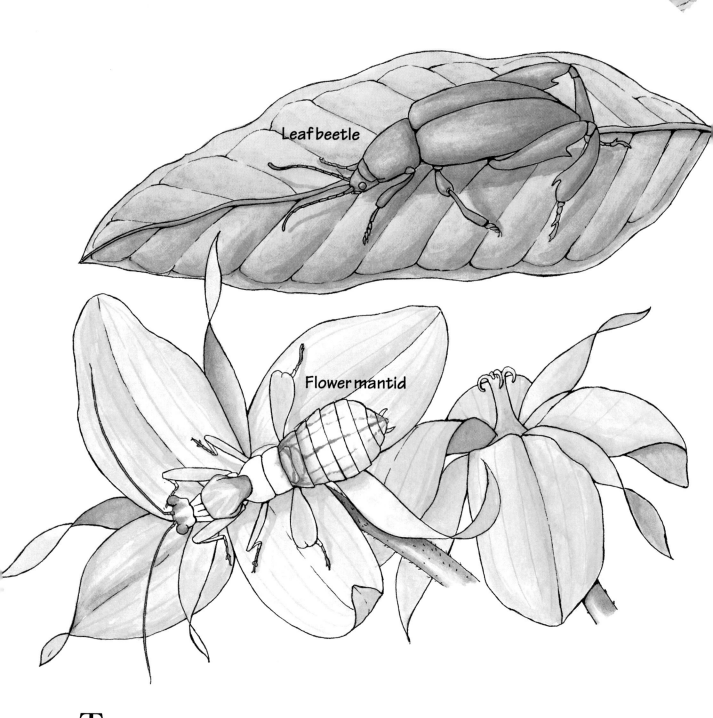

Leaf beetle

Flower mantid

The trees in the Asian rain forests are also home for many strange insects. The bright green leaf beetle shines in the light but blends so well into the dark leaves that it easily hides from its enemies. Crickets, walking sticks, long-horned and rhinoceros beetles also hide among the leaves.

18

Some insects, like the flower mantid, are the color of the leaves and flowers they live on. This camouflage hides them safely in the forest. Large atlas moths and birdwing butterflies flutter by.

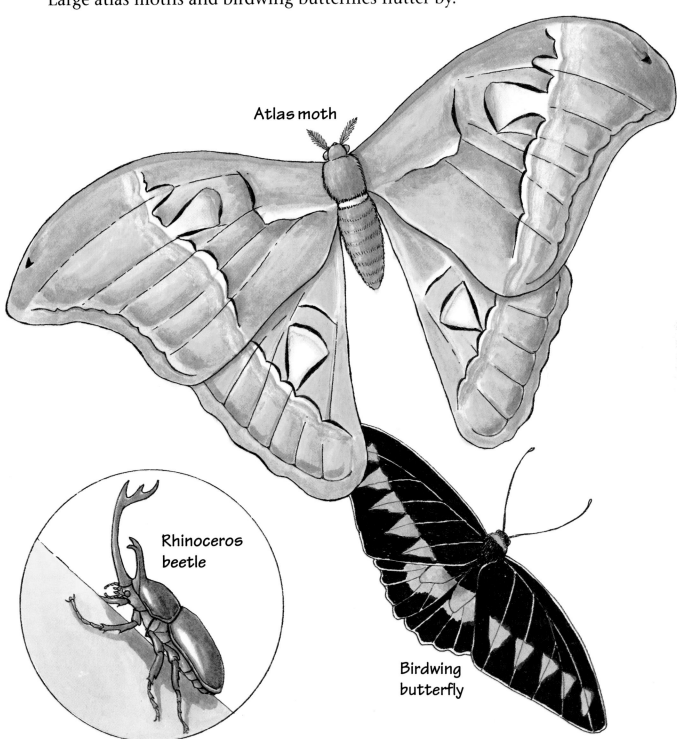

Atlas moth

Rhinoceros beetle

Birdwing butterfly

19

The tapang tree is the tallest rain forest tree in the world. Strangler figs live in this forest, too, rooting in other trees. Teak trees are treasures for their precious wood, and wild ginger is a valued spice. Up on stilted roots, the screw pine is a strange sight. Because of the shallow topsoil layer in rain forests, trees have shallow, widely spreading root systems. As a result, the trees are easier to log because they can be simply knocked over with a bulldozer. Everywhere lianas hang from trees, and ferns and flowers give the forest color.

Tapang tree

Screw pine

The pitcher plant takes advantage of the forest's many insects by trapping them in its long neck and then dissolving them in the acid at its bottom.

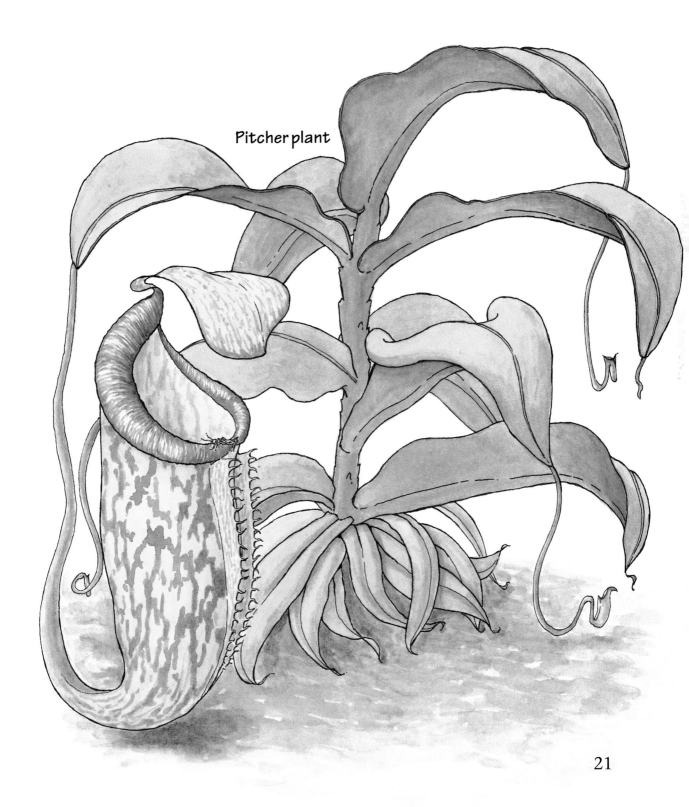

Pitcher plant

The African Rain Forest

Africa's rain forests are found mostly on the western coast and on the island of Madagascar. Chimpanzees and gorillas feed in the trees, eating leaves and fruits. Colobus, Diana, and Mangabey monkeys all live on the abundant forest leaves. The golden potto, a small primate, hunts at night for fruit and insects.

Colobus monkey

Chimpanzee

Diana monkey

Gorilla

Mangabey
monkey

Golden potto

23

Flying squirrel

Termite

Pangolin

Flying squirrels glide gently through the trees. The strange African pangolin hunts termites and ants high in the trees.

24

Genets and civets are cat-like
predators that prowl the forest floor,
while the crowned eagle hunts from
the air above. Elegant okapis and duikers
graze on stems and branches.

Crowned eagle

Genet

Okapi

Duiker

Here tailor ants sew leaves together with silken threads to build their nests. Colorful butterflies gather together in the mud on the forest floor to lap up needed minerals.

The forests are cluttered with lianas hanging from the trees. Tree ferns stand 30 feet high. Yellowwood and bamboo grow thickly here. Many rare and beautiful orchids add color to the rain forest canopy.

The Rain Forests of Madagascar

The island of Madagascar lies off the east coast of Africa. It is noted most for its strange monkey-like lemurs, though much of the plant and animal life there is unique.

The mouse lemur lives almost entirely up in the trees, leaping from branch to branch. These smallest of lemurs eat insects and plants at night, curling up in the daytime to sleep.

The ringtailed lemur is the one most commonly seen in zoos throughout the world. Unlike most lemurs that prefer to stay up in the trees, the ringtails spend much of their time on the ground traveling in groups, feeding on fruits and leaves during the day.

Mouse lemur

Ringtailed lemur

One of the strangest and least-known relatives of the lemur is the night-dwelling aye-aye. It uses its weird, long clawlike middle finger to tap tree bark in search of insects and grubs to eat.

The tiny tenrec, found only in Madagascar, searches the rain forest floor for insects to eat. It might wisely avoid some of the strange grasshoppers with their spiked armor that are found there.

The orange frog stands out against the green foliage. So do the pink flastids, feeding on the juices of plants. If they are disturbed, the flastids will drop to the ground like rose petals.

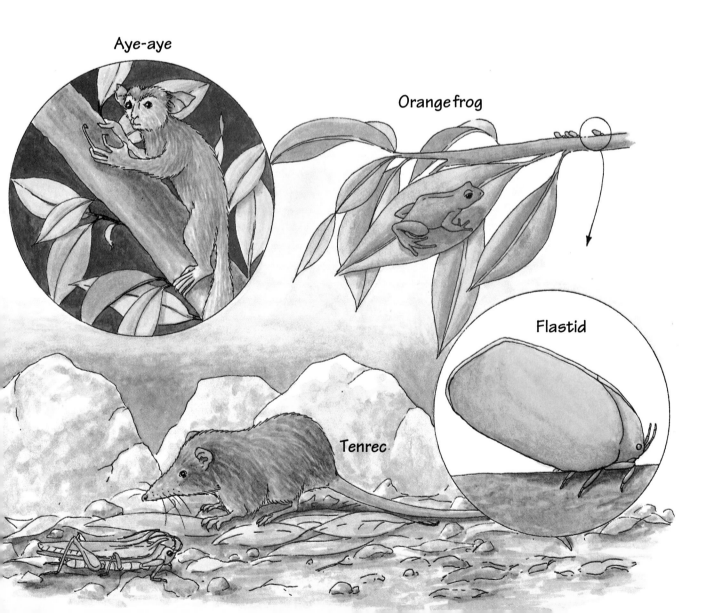

Aye-aye

Orange frog

Flastid

Tenrec

Rain Forests Today

Rain forests all over the world are being cut for timber and burned for ranch and farm land, even though the soil is so lacking in nutrients that the cleared land is good only to grow crops for about three years. Then more rain forest is cut or burned, and all its living things are lost. If these plants and animals don't exist in another part of the jungle that remains untouched, they are lost forever—some before they are even discovered!

People are just beginning to realize the importance of the rain forests. Many of the medicines found there cannot be made in the laboratory. The rosy periwinkle, found on Madagascar, has been discovered to contain a cure for a type of leukemia, a deadly disease of the blood.

Perhaps most important, as tropical rain forests absorb and release water like a great green sponge, they also absorb carbon dioxide emitted by cars and industry throughout the world. In exchange, acting like a giant filter, they send fresh oxygen into the air. Many scientists think that in this way the rain forests actually help to control the world's climate. Clearly, their destruction affects the entire earth.

Glossary

acid: a sour substance

browse: to feed on tender shoots, twigs, and leaves

camouflage: to disguise

carbon dioxide: a colorless gas formed in animal respiration and in burning substances

foliage: a mass of leaves of a tree or plant

grub: a soft, thick, wormlike larva

mineral: a solid, crystalline substance naturally occurring in the ground

nectar: a sweet plant secretion

nutrient: a nourishing substance

pollen: a mass of fine, powdery microspores of a seed plant used for fertilization

pollinate: to fertilize by spreading pollen

prey: an animal taken for food by another

sap: a watery fluid that circulates through vascular plants

spice: an aromatic product from plants used to flavor foods

Animals Index

Plants Index

574.909
AMS

Amsel, Sheri.

9582

Rain forests.

$14.93

9582

574.909   Amsel, Sheri.
AMS
         Rain forests.

WALLIS ELEMENTARY SCHOOL
WALLIS, TEXAS

RL 5.5